The Historic L
Loch Lomond

Lochearnhead and Glen Ogle, with the tree-lined course of the former Callander and Oban Railway cutting into the western slopes of the glen (NMRS D46129CN).

Royal Commission on the Ancient and
Historical Monuments of Scotland

Historic Scotland

CONTENTS

Page

3	PREFACE
4	INTRODUCTION
8	HISTORICAL DEVELOPMENT OF THE LANDSCAPE
8	The Landscape of Improvement and Industry
12	The Pre-improvement Landscape
15	The Medieval Landscape
18	The Prehistoric Landscape
22	CASE STUDIES
22	South Loch Lomond
26	Glen Finglas
28	Flanders Moss
31	CONCLUSION
32	REFERENCES

Front cover photo: Loch Lomond, viewed from the air above Balloch (NMRS D56946CN).

Back cover photo: A landscape of improvement-period fields to the north of Balloch, from a vertical aerial photograph taken during the All-Scotland Survey of 1988 (NMRS 51388194).

ISBN 1-902419-20-0
© Crown Copyright: RCAHMS and Historic Scotland 2000

PREFACE

This report has been written by Mr S D Boyle (Royal Commission on the Ancient and Historical Monuments of Scotland) and Dr L Macinnes (Historic Scotland) and edited by Dr Macinnes and Mr J B Stevenson. It is based on an analysis of existing archaeological records by Mr Boyle and Mr P McKeague combined with a Historic Landuse Assessment conducted by Ms L Dyson Bruce. Dr D J Breeze, Dr P J Dixon, Ms S Govan (Historic Scotland) Mr S P Halliday, Dr R Hingley (Historic Scotland) and Mr R J Mercer provided additional advice during the project. The maps have been prepared by Ms G L Brown. Aerial photographs were taken by Mr R M Adam and additional photographic services have been provided by Mr D Smart. Although this has been principally a desktop project, limited fieldwork was undertaken by Mr Boyle, Mr D C Cowley, Ms Dyson Bruce, Mr Halliday and Mr McKeague, with aerial reconnaissance by Mr K Brophy, Ms M M Brown, Mr Cowley and Ms Dyson Bruce. The layout of this publication was prepared by Mr J N Stevenson. Unless otherwise noted, all those mentioned above are staff of the Royal Commission on the Ancient and Historical Monuments of Scotland (RCAHMS).

Historic Scotland and RCAHMS are grateful to many individuals who assisted with various aspects of the project, including Mrs L Main, Stirling Council; Dr C Swanson and Mr H McBrien, West of Scotland Archaeology Service; Mr R Turner, The National Trust for Scotland; Ms C Ellis, AOC Archaeology Group; and the many landowners who allowed access during fieldwork, particularly Luss Estates and Mr J McNaughton, Inverlochlarig. Unless otherwise indicated, all illustrations in this report are © Crown Copyright, RCAHMS. The maps are all based upon the Ordnance Survey mapping with the permission of the Controller of the Stationery Office © Crown Copyright; licence number GD03127G/001/2000. Maps 2, 5, 6, 7, 9 and 11 incorporate data prepared by the Macaulay Land Use Research Institute.

RCAHMS was established by Royal Warrant in 1908, charged with recording all aspects of the built heritage of Scotland. RCAHMS fulfils this duty through programmes of fieldwork and through the maintenance of the National Monuments Record of Scotland (NMRS), a public archive which is open for consultation Monday to Friday from 9.30am to 4.30pm (4.00pm on Fridays). The normal public holidays are observed. Alternatively, the NMRS can be interrogated via the RCAHMS website.

Royal Commission on the Ancient and
Historical Monuments of Scotland
(National Monuments Record of Scotland)
John Sinclair House
16 Bernard Terrace
EDINBURGH
EH8 9NX
Tel: 0131-662 1456
Fax: 0131-662 1477 / 0131-662 1499
Website: www.rcahms.gov.uk

Historic Scotland is an executive agency of the Scottish Executive responsible for discharging the Scottish Ministers' functions in relation to the protection and presentation of Scotland's built heritage and for advising them on built heritage policy. It administers the Scottish Ministers' statutory duties for the scheduling and protection of ancient monuments and for the listing and protection of historic buildings. Historic Scotland offers advice and grant aid for the survey, repair and management of ancient monuments, historic buildings and historic landscapes. It also provides strategic advice to other bodies on policies for the protection of the historic environment and on associated grant schemes.

Historic Scotland
Longmore House
Salisbury Place
EDINBURGH
EH9 1SH
Tel: 0131-668 8600
Website: www.historic-scotland.gov.uk

INTRODUCTION

The area around Loch Lomond and the Trossachs has been proposed for designation as one of Scotland's first National Parks. It has been recognised that the establishment of National Parks involves a commitment to the cultural heritage of an area as well as to its natural heritage (SNH 1999). This report seeks to explore the historic landscape of the proposed National Park and to explain the important contribution that archaeology can make to the understanding of that landscape. It must be borne in mind, however, that this is no more than an interim statement, for a great deal of archaeological information in the Park remains unrecorded.

The boundaries of the Park have not been fixed at the time of writing, but this report has been guided by the proposals laid down by Scottish Natural Heritage (SNH 1999, 32-3), and includes the districts described as the 'core area' (Loch Lomond itself and the Trossachs) and as being 'potential areas for primary consideration' (Glen Falloch, Strathfillan, Glen Dochart and Glen Ogle in the north and Flanders Moss in the south). Areas designated as having potential for secondary consideration (the Endrick catchment, Glen Lochay, Loch Earn and northern Cowal) have been omitted, but many of the general points made here will also apply to them.

This report presents an analysis of the historic framework of the modern landscape, amplified by information distilled from existing archaeological records held by RCAHMS. The emphasis is on the development of the whole landscape rather than on individual sites and monuments, recognising that the modern landscape is the accumulation of history and contains elements surviving from various periods. This report attempts to highlight some of these elements and to illustrate the way the landscape of Loch Lomond and the Trossachs has been shaped and modified by human activity over time.

The report falls into four parts. This introduction includes a summary of the methodology used in preparing the report, discusses past archaeological work and looks at the current landuse of the Park. There follows an analysis of the historical development of the landscape, moving from the present landscape to its earlier phases. The third part looks at three case study areas in more detail, highlighting issues relating to particular landscapes within the Park and, finally, a conclusion draws the various threads together. A brief summary, highlighting management considerations, is provided at the end of each section.

Methodology

A Historic Landuse Assessment forms the basis of this report, combined with information from existing archaeological records. A brief discussion of the potential and limitations of these sources is given below. No field survey has been undertaken, beyond brief verification visits to a handful of individual sites.

Historic Landuse Assessment is a mechanism developed jointly by Historic Scotland and RCAHMS to map the character of the landscape, identifying both the origins of its component parts and elements of earlier, relict landscapes surviving within it (Dyson Bruce *et al.* 1999). Using aerial photographs, land cover data from the Macaulay Land Use Research Institute, Ordnance Survey maps (both up-to-date and nineteenth-century editions) and previously recorded areas of archaeological landscape, a mosaic is produced in which each piece (of at least one hectare in extent) has a Current Landscape Type and may also contain one or more Relict Landscape Types. The end product is a map, or series of maps, with a related database, which can be used to provide a broad overview of the historic landscape and an analysis of the forces of change that have acted upon it. Such an overview is extremely valuable in demonstrating the dynamic nature of the landscape and the pressures on surviving historic elements.

Archaeological records for the Park are held by the NMRS and also by Sites and Monuments Records maintained by local authorities. These records have been developed using a variety of sources, including field surveys, discoveries reported by members of the public, aerial reconnaissance, antiquarian references and historic maps. These sources vary both in the extent of their coverage and in their quality, and the resulting records are inevitably uneven.

Most of our knowledge comes from the work of the Ordnance Survey, which was active in the area during the 1960s and 1970s. Its work was invaluable in verifying and mapping reported discoveries, but its remit allowed little opportunity to prospect for unrecorded sites. In 1963 the Royal Commission published an inventory of Stirlingshire (RCAHMS 1963), which contains detailed accounts of the major monuments known at that time, and subsequently published lists of sites in the former districts of Dumbarton and Stirling (RCAHMS 1978; 1979). In the last decade a number of field surveys have been undertaken, often in response to afforestation or woodland regeneration proposals.

Ardchyle, Glen Dochart, photographed c.1900, looking towards Ben More. Note the thatched roofs held in place with timber battens (NMRS PT6654).

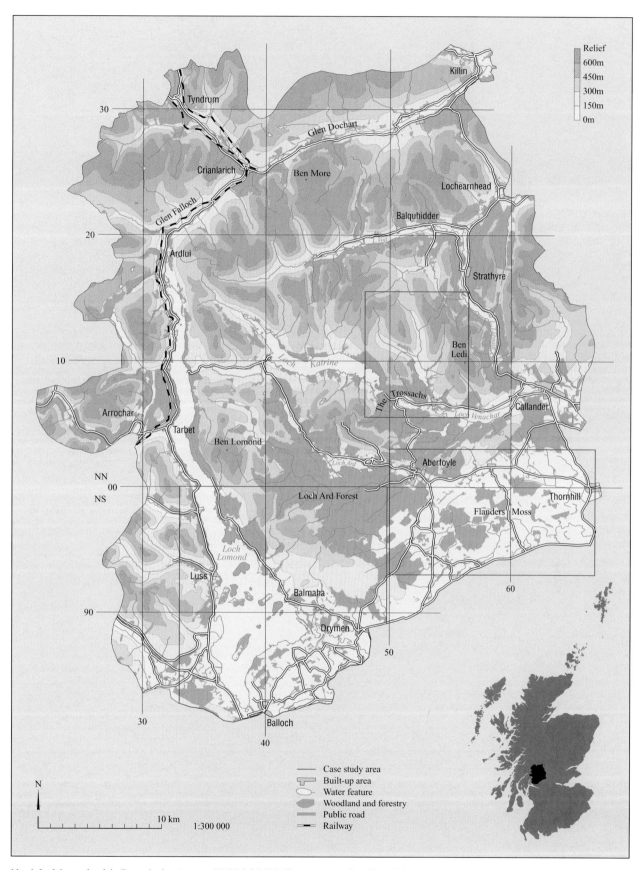

Map 1 Loch Lomond and the Trossachs: location map (NMRS DC42586). This map is reproduced from Ordnance Survey material with the permission of Ordnance Survey on behalf of the Controller of Her Majesty's Office © Crown Copyright GD03127G001/2000.

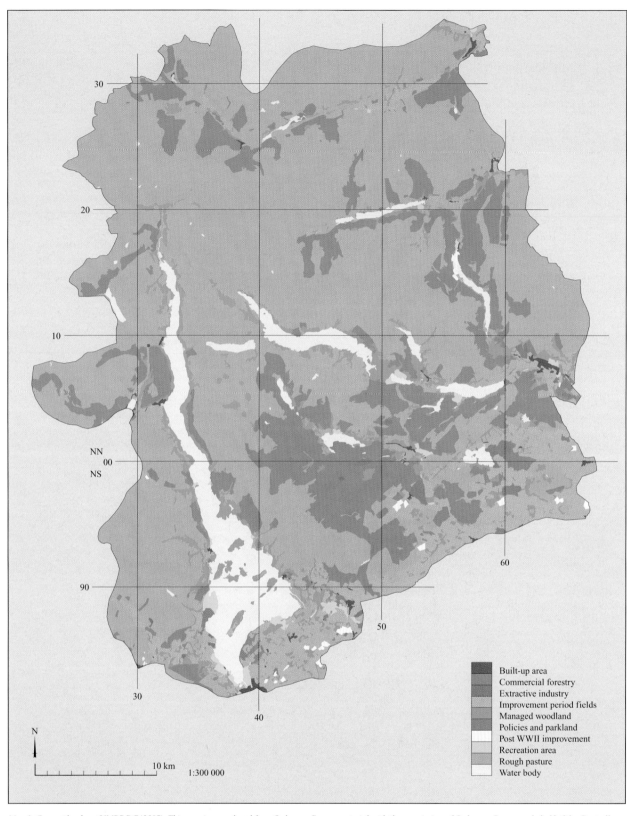

Map 2 Current landuse (NMRS DC42587). This map is reproduced from Ordnance Survey material with the permission of Ordnance Survey on behalf of the Controller of Her Majesty's Office © Crown Copyright GD03127G001/2000.

The Pendicles of Collymoon, improvement-period fields in the upper Forth valley (NMRS D56923CN).

The Lake of Menteith, with the mosses of the upper Forth valley in the foreground and the Menteith Hills beyond (NMRS D56934CN).

These have demonstrated that extensive remains, particularly of the eighteenth and nineteenth centuries, survive unrecorded in the Park (e.g. Carter and Dalland 1997; BUFAU 1997; FIRAT 1995-8; Henderson 1999), and it is now clear that many types of site are seriously under-represented in existing records, a situation that can only be resolved through further fieldwork.

If the Park has not attracted as much field survey as the quantity of remains appear to merit, there has been even less excavation. Many prehistoric sites, particularly burial cairns, were dug into in the eighteenth and nineteenth centuries, but there have been relatively few modern excavations, though some work has been undertaken, notably on the broch at Fairy Knowe, Buchlyvie (Main 1999) and at a Neolithic platform in Flanders Moss (Ellis 1999). Many of the assumptions made about the date and function of prehistoric sites within the Park are therefore based on excavations in other parts of Scotland. Similarly, there has been little attempt to explore the organisation and chronology of settlements of medieval or later date, though again there are exceptions, such as at Lix in Glen Dochart (Fairhurst 1971), Ross on the southern slopes of Ben Lomond (Hunter 1996) and Lianach in Glen Buckie (Stewart and Stewart 1989).

The Modern Landscape

Map 2 summarises the Current Landuse Types identified in Loch Lomond and the Trossachs by the Historic Landuse Assessment. The Park straddles the boundary between highland and lowland Scotland and this is reflected in modern landuse patterns. In the Highlands the bulk of the modern landuse is given over to rough pasture, managed for both sheep and game. Commercial forestry is also widespread, particularly on the southern edge of the Trossachs, where the Forestry Commission has established the Queen Elizabeth Forest Park. There are also significant pockets of broadleaf woodland, many of them along the shores of Loch Lomond; these were much more extensive until the early twentieth century (Smout 1993, 43), but since the fall in demand for oak bark and charcoal, and the closure of the Balmaha pyroligneous acid works in 1920, they have deteriorated and many have been replaced by coniferous plantations. Following current national trends, however, there are several projects within the Park aiming to increase the extent of broadleaf cover.

The two dominant upland landuse types have contrasting effects on landscapes of preceding periods. On land maintained for rough grazing and shooting there is a high potential for the survival of visible archaeological sites and relict landscapes, and in these areas field-systems predating the improvements and clearances of the late eighteenth century are commonly encountered. The planting and harvesting of coniferous plantations have had a much more destructive effect, though even here there is some potential for survival. Some archaeological features, particularly the more solidly-built stone structures, may not have been completely destroyed and some evidence could be recovered when plantations are felled. There is evidence on Loch Lomondside that oak woodlands can also survive within later forestry.

In the lowlands, most of the ground is taken up by farms whose current field-systems were laid out during the age of agricultural improvements, roughly the period 1750-1850. The resulting pattern of small rectilinear fields defined by stone dykes and hedges gives the lowland landscape much of its character. These changes altered the landscape so thoroughly that earlier landuse patterns can now only be glimpsed. Archaeological sites here tend to survive upstanding only in field corners, in shelter belts or in pockets of undisturbed land, such as wetland, though aerial photography may record the cropmarks of buried features at certain stages in the growth of wheat and barley. Now, however, the improvement-period landscape is itself coming under threat from the amalgamation of fields into larger units, and from gradual urbanisation. Commercial forestry has also made inroads into the lowlands, and it is possible that, unchecked, these forces for change will continue to erode the improvement landscape.

It was also as part of the improving movement that the policies around many large houses were established or remodelled. Many of these survive substantially intact, but here too there are growing pressures for change, and many policies have been given over to recreational purposes, particularly around Loch Lomond. In some cases the change in use may have little visual impact, but the more ambitious leisure developments may have profound effects on the historic landscape.

Summary and Management Considerations
It is clear that the modern landscape has been influenced by its use in the past, and is still subject to continual change, particularly through agriculture, afforestation and development pressures. If the historic character of the Park is to remain evident for the future, then it is important to assess the impact of these pressures on the landscape. This study aims to help clarify the historic character of the Park and highlight associated threats and opportunities.

THE HISTORICAL DEVELOPMENT OF THE LANDSCAPE

The Landscape of Improvement and Industry

The present pattern of the Scottish rural landscape was established in the late-eighteenth and nineteenth centuries, as land improvements were introduced to increase yields and profits from farming. In lowland areas the earlier landscape was transformed as new regimes of crop rotation were introduced, drainage and soil fertility were improved, fields were realigned, farm buildings were replaced and farms were amalgamated into larger, more efficient holdings. In many upland areas, however, the drive to maximise yields took a different course, as the wholesale conversion of upland townships to sheep farms rendered the existing mixed farming communities redundant, and brought about the mass depopulation of many highland glens.

Map 3 illustrates the effect of these changes on the landscape of Loch Lomond and the Trossachs. The greatest impact came from the enclosing of fields into the rectilinear units that represented order and efficiency to the improving landowners and their surveyors. In addition to improving existing farmland, new ground was brought under cultivation through drainage of the mosses, especially in the Forth valley, radically altering the local landscape. The field patterns resulting from these improvements largely define our modern image of how the countryside should appear. Map 3 depicts all improvement-period fields identified by the Historic Landuse Assessment and this can be compared with Map 2 which indicates improvement-period fields still used and maintained today. Some fields on the highland fringe have reverted to rough grazing, for example around Tarbert and Lochs Venachar and Achray. There has also been some loss of improvement fields in lowland areas, for example to the south of Aberfoyle, but, apart from loss to forestry and a certain amount of attrition through the expansion of built-up areas, there has been little change to the physical characteristics of farmland since the nineteenth century. However, future developments could alter significantly the essentially improvement-period character of the rural landscape.

It was during the improvement period that many of the villages within the Park assumed their present form. The

A landscape of improvement-period fields to the north of Balloch, from a vertical aerial photograph taken during the All-Scotland Survey of 1988 (NMRS 51388194).

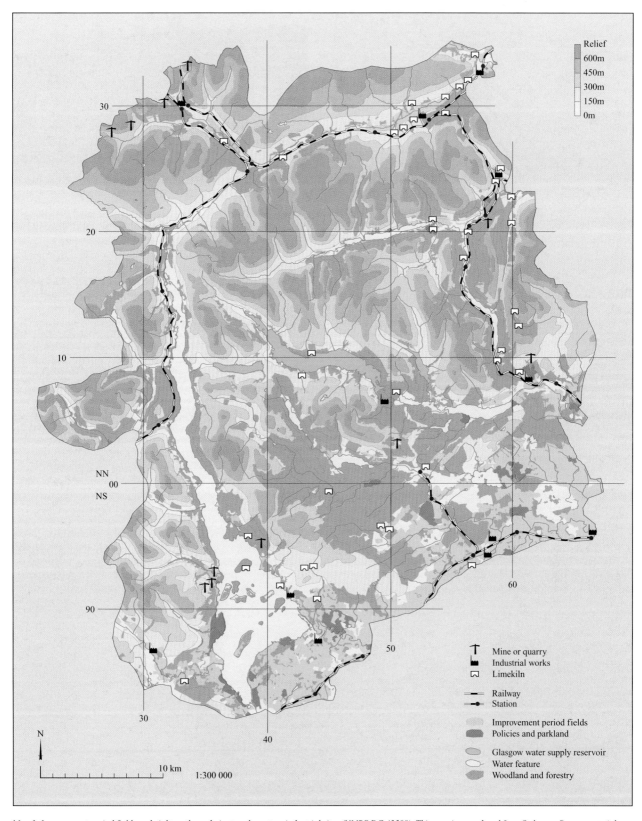

Map 3 Improvement period fields and eighteenth- and nineteenth-century industrial sites (NMRS DC 42588). This map is reproduced from Ordnance Survey material with the permission of Ordnance Survey on behalf of the Controller of Her Majesty's Office © Crown Copyright GD03127G001/2000.

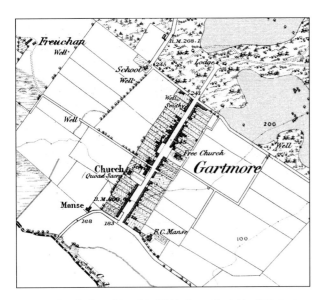

Gartmore, from the first edition of the OS 6-inch map (Perthshire 1866, sheet cxxx).

Rossdhu House, built 1772-3 for Sir James Colquhoun of Luss and now home to the Loch Lomond Golf Club (NMRS C42650CN).

fondness of improving landlords for order and regularity, displayed in the layout of their farms, extended to the design of settlements for estate workers, and several planned villages were constructed in this period. A good example is Gartmore, laid out in the eighteenth century, while the neat rows at Buchanan Smithy and Milton of Buchanan were built by the Dukes of Montrose in the early nineteenth century, and Luss was remodelled in 1850 by Sir James Colquhoun. Callander also owes its development to this period; it was originally laid out by the Duke of Perth in 1739 and, following the forfeiture of the estate after the 1745 Jacobite Rising, it was further developed as a settlement for army veterans by the Commissioners for the Annexed Estates.

As well as investing in improvements to agricultural land and workers' housing, the greater landowners of the eighteenth and nineteenth centuries further improved their estates by constructing large houses and establishing landscaped policies around them. Map 3 demonstrates that these are concentrated in the southern part of the Park, and a comparison with the distribution of castles (Map 5) shows that many of these estates had their origins in the strongholds of medieval magnates. Comparison of Maps 2 and 3 indicates that many policies have now been converted to recreational use; this is a phenomenon most clearly seen around the southern shores of Loch Lomond, whereas further east in Menteith (where there has perhaps been less demand for golf courses, hotel developments and other tourist facilities) policies have more commonly remained in private hands.

While agriculture dominated the economy of the Park in the eighteenth and nineteenth centuries, other industries also established a presence. In the eighteenth century timber became an important source of income for many highland landlords. Apart from a general demand for constructional timber, wood was consumed by the acid works at Balmaha and Aberfoyle, and bark was in great demand from the tanning industry. Charcoal was supplied as a fuel for the iron industry, and several ironworks were established in the Highlands during the eighteenth century. One of these operated in the 1720s at a site between Loch Katrine and Loch Achray, using charcoal from nearby woodlands leased from the Montrose estates. The Montrose estates, on the east side of Loch Lomond and around Aberfoyle, were particularly diligent in planting and managing woodlands during the eighteenth and nineteenth centuries. These woods were harvested in a twenty-four year rotation, with specified numbers of trees being left to grow as standards for two, three or four rotations.

Lime was in great demand, both as a source of mortar for the building industry and as an agricultural fertiliser. Limestone outcrops occur in several places to the north-west of the Highland Boundary Fault, and this is reflected by the distribution of recorded limekilns depicted on Map 3. Slate also occurs to the north-west of the fault, and there were slate

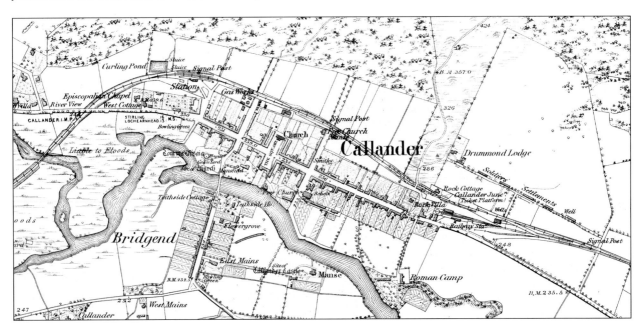

Callander, from the first edition of the OS 6-inch map (Perthshire 1866, sheets cxv & cxxiv).

quarries at Luss, Sallochy (near Rowardennan) and Aberfoyle. Farther north, around Tyndrum, deposits of gold and lead were exploited from 1739 until the early twentieth century, and the remains of a lead-crushing mill can still be seen to the south of the village. There was also a gold mine near Lochearnhead, and other industrial concerns include several textile mills, tile works and the acid works at Balmaha, which supplied the textile industry on the River Leven, immediately beyond the southern edge of the Park.

One of the greatest engineering projects of the mid-nineteenth century continues to have an impact on the landscape today – the construction of the Glasgow Corporation waterworks at Loch Katrine, completed in 1859. Loch Katrine was dammed and twenty-six miles of aqueduct were laid to a new reservoir at Milngavie. To compensate the River Teith for the loss of water, dams were also constructed on Lochs Venachar and Drunkie. A second aqueduct was added later in the century and Loch Arklet was dammed to increase the capacity of the scheme before the First World War. The latest expansion, in the 1960s, involved the flooding of Glen Finglas with a new reservoir linked to Loch Katrine.

Until the nineteenth century transport links in the area were poor. In the mid-eighteenth century some roads through the area were constructed by the military under the supervision of Major Caulfield, but these were built specifically to link highland garrisons. A few public roads were built or improved in the early nineteenth century, but many goods continued to move by water, partly to avoid payment of road tolls. Steamboats first appeared on Loch Lomond in 1818 and on Loch Katrine in 1843. By then the tourist industry was growing fast, inspired by the works of Sir Walter Scott and other romantic writers, and hotels and other facilities sprang up at the steamer piers and along the coach routes connecting them with the population centres to the south. At the north end of Loch Lomond a canal was constructed in 1847 to enable steamers to reach the Inverarnan Inn (Campbell 1999, 74-6). Tourism received an even greater boost with the arrival of railways, which reached Balloch in 1850 and Callander in 1858, before continuing north to Crianlarich (1873), Aberfoyle (1882) and Killin (1886). These communities, and other highland villages, now came within easy reach of the industrial central belt, and were transformed as they acquired hotels and villas which still provide much of their character today.

Summary and Management Considerations
The eighteenth- and nineteenth-century innovations in agriculture, industry and communications have had a lasting impact on the landscape of the Park, and account for much of its character as we see it today. In addition to the basic field pattern, there are surviving elements of settlement, associated landuse, industry and communication systems. However, this historic character could be radically altered by modern development and by the demands of agriculture and forestry. These pressures could lead to the further loss of field boundaries and the attrition of individual features and their landscape setting. One of the major challenges facing the Park Authority will be how to balance continuity from the past with change for the future.

The policies of Rossdhu House, on the west side of Loch Lomond, now a championship golf course (NMRS D56952CN).

The Pre-improvement Landscape

Before the sweeping changes of the agricultural improvements of the late-eighteenth and nineteenth centuries, agriculture in Scotland was, generally speaking, based around a system of multiple-tenancy farms, within which the houses of the inhabitants were clustered together into one or more townships. Fields were generally unenclosed, though there might be a turf or stone head-dyke around the arable ground, partly to define its extent and partly to protect growing crops from livestock. Beyond the head-dyke there was a substantial area of grazing, parts of which might also be cultivated from time to time. In upland areas, cattle were taken during the summer months to high or remote pastures known as shielings in order to make use of all available grazings, and part of the community would accompany them there, living in groups of small huts. The rural population of many highland areas appears to have reached a maximum in the late eighteenth century, but this was followed by a swift decline, caused partly by a series of poor harvests in the early nineteenth century, partly by the attraction of employment in the industries of lowland Scotland, but most of all caused by the conversion of large areas to sheep farming. To give just a few examples, between 1755 and 1801, the population of Buchanan parish fell from 1,699 to 748; between 1755 and 1831, that of Aberfoyle declined from 895 to 660; and in the same period, that of Balquhidder fell from 1,592 to 1,049.

Today, the remains of pre-improvement townships may be indicated by building foundations, ruined corn-drying kilns,

A nineteenth-century farmstead at Wester Sallochy, Rowardennan, recently cleared of trees by the Forestry Commission. Settlement remains such as these, given little attention when the forest was planted, are now regarded as valuable visitor amenities (NMRS D56227CN).

head-dykes and areas of rig cultivation, while shieling grounds are betrayed by the turf or stone footings of herdsmen's huts. It is not usually possible to date such remains with any precision, but it is likely that most of the visible evidence relates to the final flourish of pre-improvement agriculture in the eighteenth century. Within individual townships it is sometimes possible to identify differences between buildings or episodes of rebuilding that suggest there was some chronological depth to the settlement. Indeed, it is possible that some settlements may have originated in the medieval period, though it is usually difficult to demonstrate this from the surface remains alone, albeit that there is no doubt that there was extensive settlement throughout the Highlands at this time.

The recorded remains of pre-improvement settlement and agriculture are depicted on Map 4. It is apparent from the map that the visible evidence for settlements of this period is almost entirely concentrated in upland areas, and this demonstrates the thoroughness with which the lowland landscape was remodelled during the improvement period

The stone footings of a cluster of shieling-huts on a hillside above the reservoir in Glen Finglas. Small groups of buildings like this can be found at former summer grazings in the upper reaches of many highland glens (NMRS D28138).

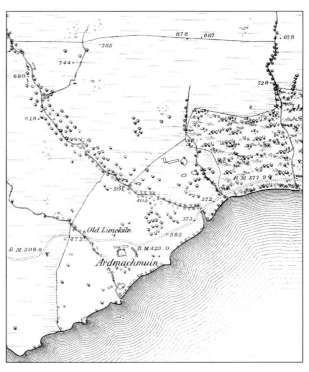
The ruined township of Ardmachmuin, on the north shore of Loch Katrine, depicted on the first edition of the OS 6-inch map (Perthshire 1867, sheet cxiii).

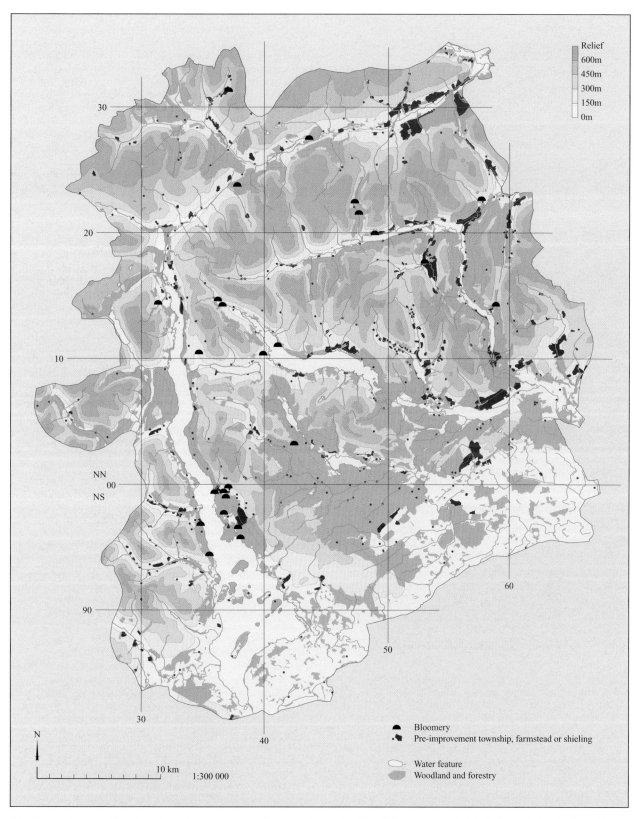

Map 4 Pre-improvement settlement and bloomeries (NMRS DC 42589). This map is reproduced from Ordnance Survey material with the permission of Ordnance Survey on behalf of the Controller of Her Majesty's Office © Crown Copyright GD03127G001/2000.

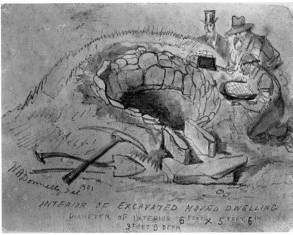

Two contemporary illustrations of shieling-huts on the Auchengaich Burn in Glen Fruin, drawn during excavation in 1901 (Top, NMRS A41743CN & bottom, NMRS A41741CN).

(compare Maps 3 and 4). Post-improvement sheep farming and deer stalking, by contrast, have posed little threat to the historic landscape in the uplands, and away from modern farms and settlements survival of landscape features is often good. However, in the late twentieth century much has been lost to afforestation schemes. In Map 4, it is striking how extensive are the remains of pre-improvement settlement in the gaps between forestry plantations in the major glens, and it seems reasonable to assume that settlement and cultivation originally spread in an almost continuous band along the lower slopes of some of these glens. This assumption is supported by the evidence from mid-nineteenth-century Ordnance Survey maps, which show many abandoned settlements in areas now afforested. Many of these will have been destroyed, but some stone structures may survive within the trees, especially if they were not subjected to deep ploughing before planting. It may well be possible to locate some of these when the forests are clear-felled.

Map 4 also shows the known sites of bloomeries – small-scale ironworking furnaces, identifiable today from their waste heaps of slag and charcoal. Twenty-five bloomery mounds have so far been identified, but they are inconspicuous monuments, easily missed, and most of those shown on the map were discovered as a result of disturbance during forestry operations. None of the sites in the Park have been excavated, but radiocarbon dates from examples elsewhere in Scotland indicate a medieval date for their use. This would seem to agree with documentary evidence for iron-working in the Loch Katrine area in the mid-fifteenth century (Aitken 1973, 194), though bloomeries may have had a long currency, and small-scale works of this type probably remained in use until the establishment of coke-fired furnaces in the late eighteenth century. Some associated charcoal-burning platforms have been located in Loch Lomondside and Strathyre, but there are likely to be many more as yet unrecorded. As the iron industry demanded large quantities of timber, it is clear that woodland management must have been important in this period, and some of the relict oak woodland in the Park may reflect this.

Summary and Management Considerations
Remains of this period are often well-preserved and can seem extensive, but the Historic Landuse Assessment demonstrates that they have been largely obliterated in the lowland area as a result of later ploughing, and have been considerably reduced in upland areas through afforestation. While some structural remains will survive in reasonable condition within forestry plantations, associated field-systems are unlikely to have survived. The surviving remains are highly vulnerable to damage through further afforestation, including restocking, or through re-development. It is often possible to plan new plantations in such a way that specific archaeological features can be protected, while their re-use can be controlled through the planning process. At the same time, however, attention should be paid to the setting of settlements within the wider landscape, as this reflects their relationship to each other and to the environment they exploited. This requires a landscape-based and integrated approach to conservation which should be possible within the National Park.

Remains of this period lend themselves well to access and interpretation for recreation and education. The various components of townships and their associated field-systems are often visible, evoking a past landscape and a way of life which, though long vanished, is nevertheless familiar to us today.

The Medieval Landscape

At present, it is not easy to gain an appreciation of the wider landscape of the medieval period through the surviving evidence, though it is likely that the pre-improvement landscape contains within it earlier patterns of use. Nevertheless, a variety of archaeological remains point to the development and use of the landscape in this period (Map 5).

For the medieval period, sites of secular and religious authority dominate the record, while rural settlements and associated landuse are harder to locate. By about 600 AD, the mountains around Loch Lomond appear to have formed a frontier zone between the Britons of Strathclyde to the south, the Scots of Dalriada to the west and the Picts to the east. No settlement remains of this date have yet been found within the Park, but it is likely that in the often turbulent conditions of the time some of the earlier forts, homesteads and crannogs (see Map 6) continued to be occupied. Further instability may have been introduced by the onset of Viking raids in the ninth century. In 870 Alt Clut, the stronghold of the Britons on Dumbarton Rock, was besieged and sacked by Vikings, but they appear to have had little lasting impact on the Park, their only appearance in the archaeological record perhaps represented by a sword, spearhead and shield boss recovered in 1851 from a cairn at Boiden on the south-west shore of Loch Lomond.

Castles first appear in the landscape during the twelfth century. The earliest were built of wood and set on earthen mounds known as mottes; four have been recorded in the Park, the best preserved being Catter Law south of Drymen. Mottes are associated with an influx of Anglo-Norman knights who settled in many parts of central and southern Scotland under royal patronage in the twelfth and thirteenth centuries. Moated sites, rectangular earthworks defined by a ditch, are grouped with mottes on Map 5, though they are slightly later in date (c.1250-1350) and were apparently built with defence less of a priority, perhaps serving as estate centres. What both groups have in common, though, is a lowland distribution, concentrated within the earldoms of Menteith and Lennox in the Forth and Endrick valleys. The location of these sites in agricultural land, together with their earthen construction, makes them particularly vulnerable to erosion, both through natural processes and through agricultural activity.

Of the thirty stone castles and towers depicted on Map 5, only three are occupied today, though parts of at least three others have been incorporated into later buildings and others survive as roofless shells. With few exceptions (such as the fourteenth-century castle on Inchmurrin) most of the surviving castles are small tower-houses of fifteenth-century date or later, though documentary evidence and local traditions suggest that at least some of these towers replaced earlier fortifications, an important example being Balloch Castle, built in 1238 for the Earl of Lennox and finally replaced by the present house in 1808.

About half of the known castles are concentrated in the more fertile lowlands in the south of the area. Here many castles were the foci of estates and retained their importance into modern times. In some cases the castle or tower has been subsumed into a more substantial mansion (e.g. Cardross and possibly Catter House), in others the tower has been abandoned for a later building on an adjacent site (e.g. Gartartan, Rossdhu and Buchanan Castle). Whether the medieval building has remained in use or not, it is striking that about half of the lowland castle sites stand within policies of eighteenth- or nineteenth-century date. Thus there has been considerable continuity in the occupation of these sites despite changing architectural fashions.

In upland areas, several towers were given additional strength by being sited on small islands. Some islands show no evidence of fortification, but are well attested as clan strongholds, such as Eilean Molach on Loch Katrine (MacGregor) and Eilan Rowan at Killin (MacNab). The

The ruins of fifteenth-century Rossdhu Castle, seat of the Colquhouns of Luss until Rossdhu House was built next to it in the 1770s (NMRS C42701CN).

The ruins of Gartartan Castle stand in the grounds of Gartmore House, on the edge of the Forth valley. The castle walls were extensively robbed to provide stone for a walled garden in the eighteenth century (NMRS C7253).

lowland Clairinsh, seat of the Buchanans, also falls into this category. As the need for defence subsided so did the advantages of an island situation, and these sites may have become something of an inconvenience, leading to their eventual abandonment. Today, only two upland castle sites remain in use at all – Duchray and Edinample, sixteenth-century towers that are both still occupied. Ironically, the abandonment of the more inaccessible castles has probably helped secure the preservation of their ruins, which have not become convenient stone quarries for later builders, a fate that has befallen so many lowland sites. However, many of these sites are overgrown with vegetation, while surviving masonry is gradually being lost through natural decay.

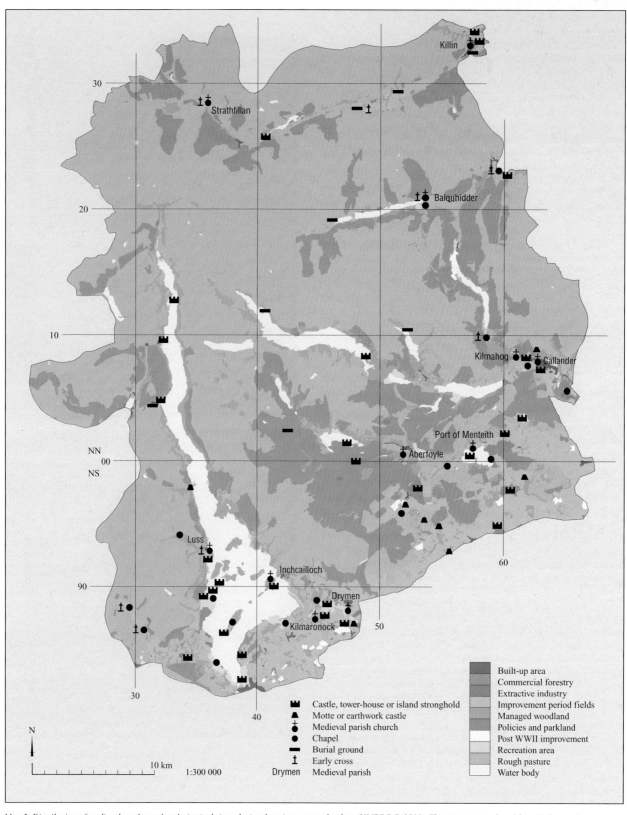

Map 5 Distribution of medieval castles and ecclesiastical sites, depicted against current landuse (NMRS DC42590). This map is reproduced from Ordnance Survey material with the permission of Ordnance Survey on behalf of the Controller of Her Majesty's Office © Crown Copyright GD03127G001/2000.

Christianity probably arrived on Loch Lomond in the sixth century AD. Many places have traditional associations with early saints. While some of these traditions may have originated during a revival of interest in Celtic saints in the later Middle Ages, there are strong associations of St Kessog with Luss, of St Fillan with Strathfillan and of St Kentigerna with Inchcailloch. No visible remains survive of any religious building dating to this period, but the early cross-slabs found at nine sites and the Norse style hog-backed gravestone of tenth-century date at Luss indicate the sites of at least some early foundations.

By about 1300, most of the parish churches illustrated on Map 5 had been established. It is immediately noticeable that churches and castles were often situated close together, emphasising the close relationship between the Church and secular authority. Most of the church sites are still in use, though the churches themselves have all been rebuilt. Indeed, at only three sites – Inchcailloch, Strathfillan Priory and Inchmahome Priory (Port of Menteith parish) – are there any visible remains of medieval date, though medieval graveslabs have been recorded in the churchyards at Killin, Balquhidder, Luss and Kilmaronock. There are also records of at least sixteen chapels in the area (that is, small churches lacking the 'cure of souls' and other administrative duties accorded to parish churches), six of which survive as ruins today. Most chapels were apparently abandoned following the Reformation, though burial-grounds attached to some of them continued in use at least until the eighteenth century. In some parts, particularly in highland glens remote from any parish church, there are burial-grounds that appear never to have had a church or chapel attached. These include several traditional clan burial-grounds, such as the MacGregor burial-ground at East Portnellan on Loch Katrine, moved in 1922 to prevent it being submerged beneath the reservoir, and two burial-grounds associated with the MacNabs, one in Glen Dochart, the other at Killin. Map 5 depicts eight burial-grounds that may have medieval origins.

Summary and Management Considerations

Unlike later periods, it is hard to detect typical rural settlements of the medieval period. It seems likely that at least some of the pre-improvement settlements and their associated landscape elements (Map 4) had their origins in this period, and further survey, detailed archaeological investigation and examination of documentary evidence will undoubtedly provide additional clues here. It is also likely that some settlement types that have their roots in the prehistoric period, particularly forts, homesteads and crannogs, continued to be built, or were re-used, in the medieval period. Similarly, some ploughed-down sites revealed as cropmarks may prove to be of medieval date.

Remains of this period are highly vulnerable to damage and decay, both through natural processes and through human activity. Those in lowland locations can be subject to erosion of earthworks, and suffer damage through farming activities or modern development. In upland areas afforestation and natural regeneration of trees and other vegetation are particular problems for fragile remains. However, it is possible to address these problems, through, for instance, consolidation of stonework or repair of erosion scars. Damage through human activity can be minimised by careful planning and an integrated approach to land management.

Features of this period often have a dramatic impact within the landscape, and can offer attractive opportunities for tourism and recreation within the Park. At the same time, though, the sensitivity of the remains to pressures from visitors and other recreational activities, including those on the lochs, must be borne in mind.

The ruins of Inchmahome Priory, on the Lake of Menteith, now in the care of Historic Scotland (© Historic Scotland).

The Prehistoric Landscape

As for the medieval period, it is not easy at present to appreciate the prehistoric landscape of the Park, though, once again, there is a variety of archaeological evidence to guide further work. Map 6 depicts the known prehistoric sites, which include burial cairns, standing stones, stone circles, cup-marked rocks, forts, crannogs, homesteads and findspots of a variety of stone, bronze and iron tools and weapons. Settlement evidence is concentrated in the southern lowlands, though there are a number of sites and findspots in several highland glens. It is likely that this distribution reflects, in part, the higher quality of the land in the lowlands, but it also reflects the increased likelihood of the discovery of artefacts on agricultural land, where the surface is repeatedly disturbed by ploughing.

The earliest inhabitants of the Park arrived by about 7000 BC. These were groups of semi-nomadic hunter-gatherers, whose presence is known from occasional chance discoveries of worked flint and red deer antler, perhaps lost during hunting expeditions. No settlements of this period are known in the area, though a group of artefacts from Glen Finlas on the west side of Loch Lomond may indicate the site of an encampment.

In the succeeding Neolithic period, beginning about 4000 BC, a more settled way of life became established, characterised by the cultivation of cereals, animal husbandry and a new range of tools and artefacts, including, for the first time, pottery. There is little trace of the settlements of these first farmers in the Park, though cropmarks of a large subrectangular structure identified in a field to the south of Callander may mark the site of a timber house of this period, while a timber platform excavated in peat on the edge of Flanders Moss may have been a jetty used for expeditions into the marshland of the upper Forth valley. Certainly, stray finds of polished stone axes, a characteristic tool of the period, demonstrate widespread Neolithic activity in the area. Rather more is known about the funerary and ritual practices of this period than about settlement. In the Park there are several cairns containing the ruins of stone burial chambers, while the stone circle at Killin may be of late Neolithic date. Far more numerous than these sites, however, are cup-and-ring marked rocks, many of which probably belong to this period. These are widely distributed, with a particularly large concentration of at least fifty carved outcrops and boulders on moorland to the north of the Lake of Menteith.

This Neolithic chambered cairn at Auchenlaich, Callander, measures over 320m in length – the longest burial mound in Scotland. The cairn now lies between a caravan park and a quarry (NMRS D59075CN).

The burial chamber of a Neolithic cairn at Edinchip, near Lochearnhead (NMRS A79852CS; © C Appleby).

A Neolithic timber platform in Flanders Moss, during excavation in 1999 (© Historic Scotland).

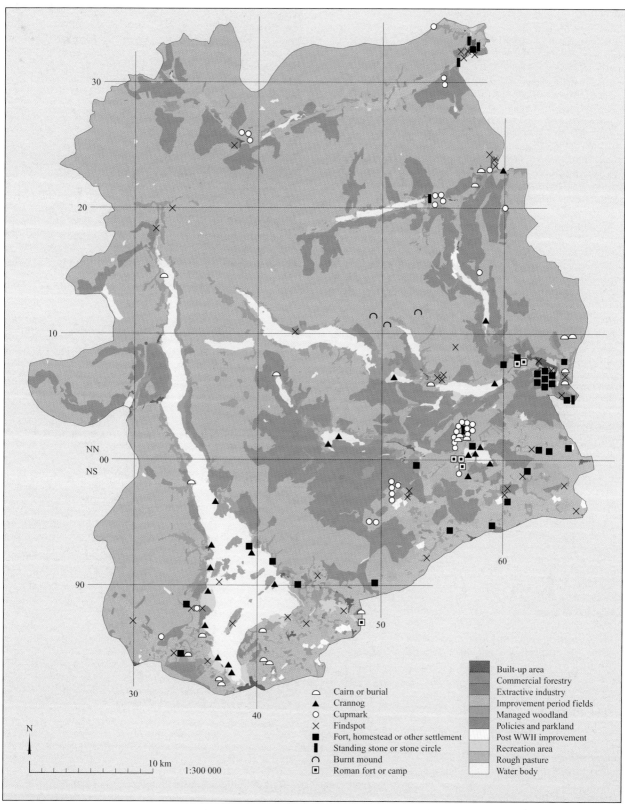

Map 6 Distribution of prehistoric and Roman monuments, depicted against current landuse (NMRS DC42591). This map is reproduced from Ordnance Survey material with the permission of Ordnance Survey on behalf of the Controller of Her Majesty's Office © Crown Copyright GD03127G001/2000.

For the Bronze Age, from the late third millennium until about 700 BC, we are again dependent largely on burial monuments (usually round cairns) and stray finds of bronze artefacts for our information on human activity within the Park. Sixteen cairns and other burials of this date have so far been recorded; half of them are situated in the lowland landscape around the south end of Loch Lomond, while the others are more widely dispersed across the area, with one group of three cairns on moorland to the north of the Lake of Menteith. Once again, little evidence of contemporary settlement has so far been recorded in the Park.

Settlements of later prehistory, by comparison, are relatively common, and a wide variety of types is represented on Map 6, including forts, duns, brochs, homesteads and crannogs. About fifty sites have been recorded to date, and there is good reason to think that others remain to be discovered, through both ground survey in the uplands and aerial reconnaissance in the arable lowlands.

Most of the ten known forts were defended by earthen ramparts fronted by ditches, though two, Strathcashell Point on Loch Lomond and Dunmore above Callander, have stone walls. Three duns, essentially small thick-walled forts of a type particularly numerous in the West Highlands, have been found. Limited excavations at one of them, Shemore on the west side of Loch Lomond, produced radiocarbon dates indicating occupation in the mid- to late first millennium BC. The brochs (one at Craigievern, near Drymen, the other at Auchinsalt, in Menteith) are another type of settlement more commonly found in the north and west of Scotland. Probably dating to the beginning of the first millennium AD, these circular towers, defended by massively-thick stone walls, form part of a wider group of such structures in southern Scotland.

The various types of defended settlement probably span a period of almost a millennium. The reasons for the variety of forms, and the social and political circumstances that led to their construction, are not yet understood. One thing that most of them have in common is their location within the landscape, usually at the edge of cultivated ground in the most fertile parts of the area, on low hills or promontories. Today they often lie within woodland, protected from the more zealous agricultural regimes of the last two centuries.

In the two remaining types of settlement, defence appears to have been less of an over-riding consideration. At least twelve circular or oval timber homesteads have so far been identified from cropmarks on aerial photographs. They are distributed from Balmaha in the west to Callander in the east and include a remarkable group around The Clash, to the south of Callander, where at least six sites have been recognised. None of these sites have been excavated but evidence from further afield has demonstrated that they represent the remains of farming settlements of the first millennium BC. Similar sites may well have existed on higher ground now given over to permanent pasture, but here the chances of their discovery are very slim.

The final group of sites that can be attributed to the later prehistoric period are crannogs, of which over twenty are known in the Park. These artificial islands, built of boulders and timber piles, provided a solid platform for timber round-houses, and their situation provided a degree of security as well as immediate access to a wide range of resources. Available radiocarbon dates suggest that crannogs began to appear around 500 BC, and continued to be used for settlement into the early medieval period. Some lasted longer still, like Elan Rossdhu, a crannog upon which stood a castle which remained the seat of the Colquhouns of Luss until at least the mid-fifteenth century. These sites have great potential for the recovery of organic remains, which do not survive elsewhere, and recent underwater survey does indeed suggest that timber has survived on some sites (FIRAT 1995-8).

The Romans make a brief appearance in the archaeology of the Park. Three Roman forts have been recorded, Drumquhassle to the east of Drymen, Malling at the west end

Dunmore, a prehistoric fort perched on a hill overlooking Callander (NMRS D24778CN).

Prehistoric settlements revealed as cropmarks near Callander. In the centre of the photograph three intersecting rings (A-C) mark the lines of timber palisades around a sequence of Iron Age homesteads. Above, there is a building (D) containing rows of post-holes, possibly of Neolithic date (NMRS PT5524).

of the Lake of Menteith and Bochastle to the west of Callander. These were constructed in the late first century AD following the Roman victory at Mons Graupius. All were carefully positioned at the mouths of highland glens. Beside the fort at Bochastle there is a temporary camp, and two more lie next to Malling: these may have been labour camps erected to house the troops building the forts. Parts of the ramparts can still be seen at Bochastle and Malling, but other sites survive only as cropmarks.

'The Kitchen', a prehistoric crannog on Loch Lomond, close to Balmaha (NMRS D56228CN).

Summary and Management Considerations

Despite only limited field survey, it is evident that there was prehistoric settlement throughout the Park. In lowland areas, much information has been lost through later development and agricultural activities, though, paradoxically, this has itself sometimes led to the recovery of some information. In these areas sites survive as isolated features or as cropmarks, invisible on the ground surface and vulnerable to modern agricultural activities and to development pressures. In upland areas, timber sites cannot easily be traced, but a range of stone-built features survive outwith forestry plantations and are clearly vulnerable to further afforestation. Within existing plantations, there is the potential for some recovery of information when trees are felled, though the remains may be too slight to be readily visible in disturbed ground. Recent work has clearly demonstrated the value of areas of peat, wetlands and lochs in preserving archaeological and palaeoenvironmental evidence which has been lost elsewhere. These are highly vulnerable to damage and desiccation and are accordingly a high priority for conservation.

Recent work has also shown that modern survey, both from the ground and from the air, excavation techniques and palaeoenvironmental research can recover new sites and enhance our understanding of the prehistory of the Park. The opportunities for recreation and tourism, and education, are correspondingly immense, though, once more, the fragility of many of these remains must be remembered.

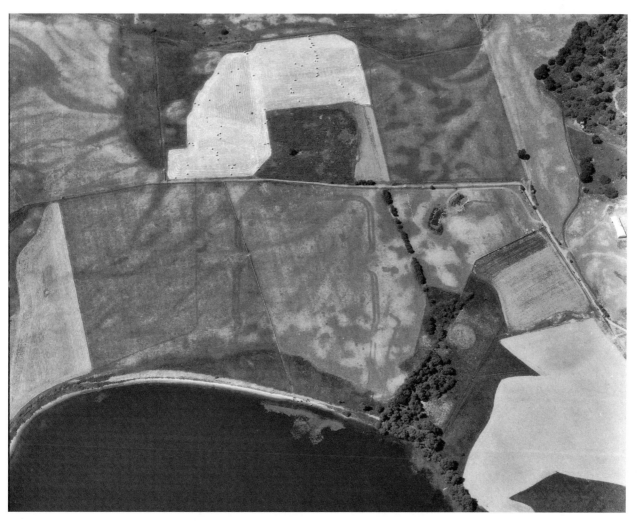

The defensive ditches and the inturned entrances of the Roman fort at Malling show up clearly as cropmarks. To the left of the fort there can be seen one corner of a temporary camp, possibly used to house troops building the fort (NMRS PT5515).

CASE STUDIES

South Loch Lomond

The terrain around the southern end of Loch Lomond can be divided into two very different areas, roughly along the line of the Highland Boundary Fault, which runs from south-west to north-east across the loch, through the islands of Inchmurrin and Inchcailloch. To the north, mountainous landscapes rise steeply from the loch shore, while to the south there is a more gentle, rolling countryside. This division is reflected in both current and relict landuse patterns, and also in the archaeology (Maps 7 and 8).

Most of the ground to the north of the fault is given over to either rough grazing or commercial forestry. There is also a significant amount of broadleaf woodland, but there are only small pockets of improved farmland. The two dominant landuse types have very different implications for archaeological remains. Sheep farming generally poses little threat to the remains of earlier periods, and some extensive areas of pre-improvement settlement survive here, particularly on the east side of the loch. By contrast, large-scale coniferous afforestation can be extremely destructive, and there can be little doubt that much archaeological evidence has been lost without record in these areas. Nevertheless, it is possible that some sites in dense plantations may not have been completely lost. Indeed, it is often only during forestry operations that some sites are recognised at all, for example bloomery mounds, most of which were first discovered by forestry workers.

The broadleaf woodlands were once of considerable economic importance, with markets in the construction and tanning industries as well as the acid factory at Balmaha, which by the late nineteenth century was consuming all the timber from the Rowardennan area (Campbell 1999, 146). Now they are managed more for their value to the landscape or to wildlife conservation, but there is a growing awareness that they also have an archaeological value, as relict artefacts of former landuse regimes (e.g. Smout 1997). It is possible that earlier monuments survive within such woodland, as ploughing was not undertaken prior to planting.

In the more gentle ground around the southern tip of the loch, in lower Strathendrick and in Glen Fruin, the landscape is characterised by rectilinear fields of the improvement period (Map 7). Those improvements have left little of the agricultural landscapes that preceded them, and the archaeological evidence for earlier periods is largely confined to isolated monuments and stray finds of artefacts (Map 8). These monuments include several Neolithic and Bronze Age burial cairns, which demonstrate that the fertile ground at this end of the loch has attracted settlers since early prehistory. Most of the recorded stray finds are also of Neolithic or Bronze Age date; they include stone axes, hammers and arrowheads. The majority of these have been discovered during ploughing, drainage works or other agricultural operations, so it is not surprising that most were found in the areas of improved fields. There are likely to be many undiscovered artefacts and sites concealed in areas of unimproved ground too, though there is less chance of these being retrieved. Settlement also extended onto the loch itself in the form of crannogs, and, in the medieval period at least, islands such as Inchmurrin, Inchcailloch and Clairinsh were also inhabited.

Around the shores of the loch a number of extensive policies were developed in the eighteenth and nineteenth centuries around the houses of the principal landowners, and this trend continued in the late nineteenth century as wealthy Glasgow industrialists built houses on the loch. A lack of public access to the shores of the loch prompted Glasgow Corporation to purchase Balloch Castle in 1915 and open its grounds to the public (Campbell 1999, 38). In more recent decades a large number of other houses and policies have been converted to recreational use, particularly as hotel developments and golf courses. The extent of this change in use can be seen from a comparison of the recreation areas on Map 2 with the policies shown on Map 3.

One important resource whose exploitation has left little trace on the archaeological record is the loch itself, and in

The island of Clairinsh, Loch Lomond, was once a seat of Clan Buchanan. Off the northern tip of the island (bottom right in the photograph) there is a crannog known as 'The Kitchen' (NMRS D46100CN).

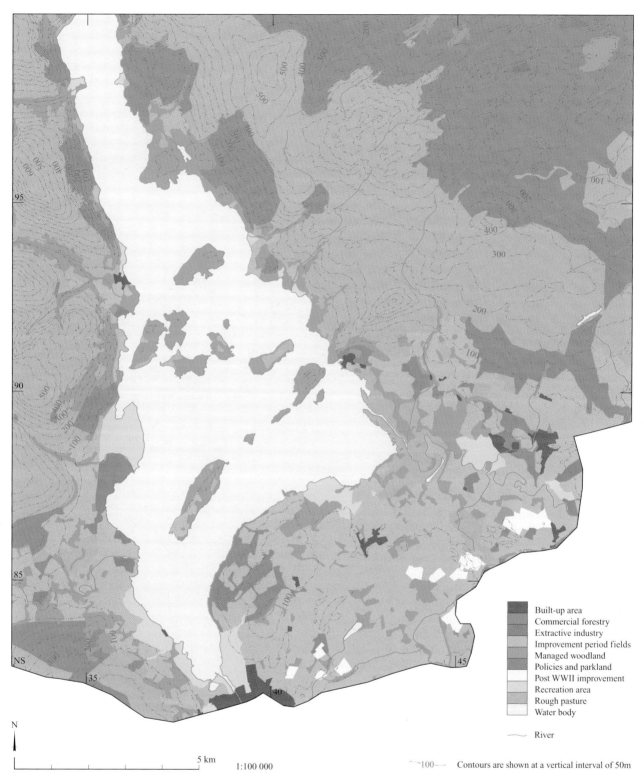

Map 7 South Loch Lomond: current landuse (NMRS DC42592). This map is reproduced from Ordnance Survey material with the permission of Ordnance Survey on behalf of the Controller of Her Majesty's Office © Crown Copyright GD03127G001/2000.

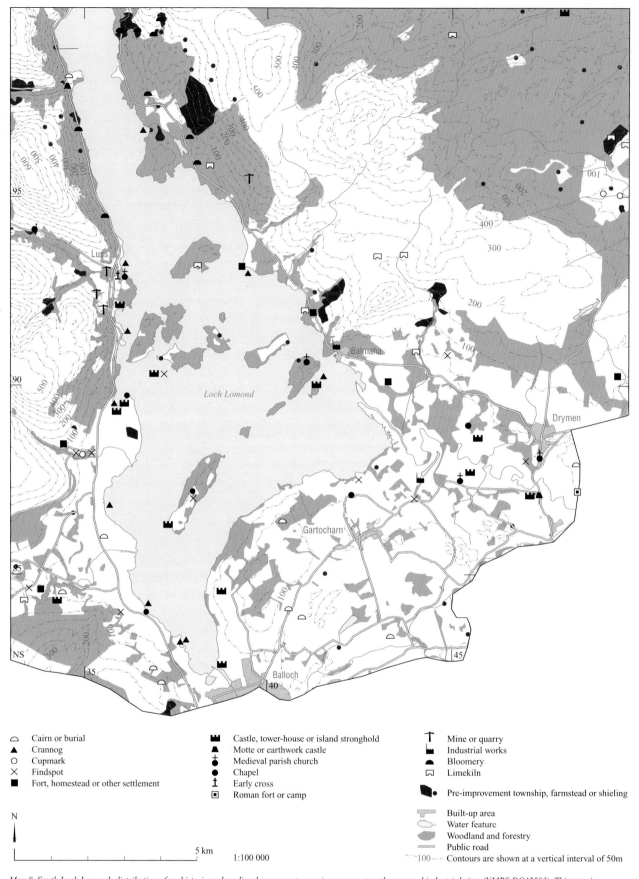

Map 8 South Loch Lomond: distribution of prehistoric and medieval monuments, pre-improvement settlement and industrial sites (NMRS DC42593). This map is reproduced from Ordnance Survey material with the permission of Ordnance Survey on behalf of the Controller of Her Majesty's Office © Crown Copyright GD03127G001/2000.

The ruins of Buchanan Castle, a nineteenth century mansion built for the Duke of Montrose, are now hemmed in by modern development (NMRS D46095CN).

Buchanan Castle, as it was envisaged by its architect, William Burn, in 1852 (© RIBA Library Drawings Collection; NMRS STD7221P).

particular its use for transport in periods when road communications were poor. The River Leven provided a route to markets on the Clyde and beyond, and many industries exploited this to the full. Slates from Luss and Sallochy, timber from many woodlands, acid from Balmaha, sandstone flags from the quarries at Kilmaronock and many other goods were shipped across the loch, while a ferry was maintained between Rowardennan and Inverbeg, and others linked islands to the mainland. There is little recorded archaeological evidence of this traffic, apart from piers at various points around the loch, and most of these are relatively late constructions, built for the nineteenth-century tourist traffic.

Summary and Management Considerations
This case study highlights the long cultural history of the loch and the impact that the development of industry and modern communications had at its southern end. Relict communication networks, such as railways and steamer routes, might provide useful opportunities to develop the tourist potential of this area, though it is important to consider the impact of modern recreational pressures on the historical character, both on the loch itself, where the crannogs are especially vulnerable, and on the surrounding land.

Glen Finglas

Glen Finglas is an upland glen between Loch Katrine to the west and Ben Ledi to the east. It is about 10km long, and with its subsidiary glens, Gleann nam Meann and Gleann Casaig, it has a catchment area of about 49 square kilometres. Most of the ground is steep-sided moorland, given over to rough grazing, but there are several significant areas of oak woodland as well as extensive coniferous plantations on the west side of the glen (Map 9). Much of the floor of the glen was flooded in 1965 when a dam was constructed to increase the capacity of the Glasgow water supply. A programme of woodland regeneration has recently started, and in preparation for this an archaeological survey was commissioned in 1996 (Carter and Dalland 1997). The area of the survey is outlined on Maps 9 and 10.

Before survey, there was only one archaeological site recorded, a possible medieval burial-ground known as Cladh nan Casan. During fieldwork, about 250 structures and areas of former cultivation were mapped, all dating to before the mid-nineteenth century (Map 10). These include, on the lower slopes, rectangular buildings grouped into farmsteads and townships, with associated enclosures and fields of rig cultivation; on higher ground there are groups of shieling-huts, comprising as many as nineteen huts each. While most of the buildings were constructed of stone, others appear to have been built of turf, suggesting that the remains are not all contemporary, but span a considerable period of time. Dating individual structures is often difficult, especially without excavation. It is likely that most of them are of eighteenth- or early nineteenth-century date, but some may date back to the medieval period, when the glen was part of a royal hunting forest. Prehistoric activity in the glen was also demonstrated by the discovery of two burnt mounds – cooking sites probably of Bronze Age date – and a third burnt mound has been discovered since the completion of the survey.

The 1996 Glen Finglas survey underlines the indispensable role of ground survey in recording archaeological remains of all periods. As a direct result of it the glen stands out on the general map of pre-improvement settlement in Loch Lomond and the Trossachs (Map 4) as an area apparently more densely populated than most of the glens around it, even though much of the most suitable ground for settlement lay beyond the reach of the survey, submerged by the reservoir. It can safely be assumed that intensive systematic surveys in neighbouring glens will fill out the settlement pattern to match that recorded in Glen Finglas.

Summary and Management Considerations
The survey of Glen Finglas demonstrates the important contribution archaeological survey can have on our understanding of the development of a landscape. This understanding can inform decisions about the future management of the landscape, as well as conservation strategies for the cultural heritage itself. The Glen Finglas area itself offers a further opportunity for access, interpretation and education.

Low winter sunlight highlights the corrugations of pre-improvement rig cultivation around a modern hill farm in Glen Finglas. Most of the buildings associated with these fields now lie beneath the waters of the reservoir (NMRS D24791CN).

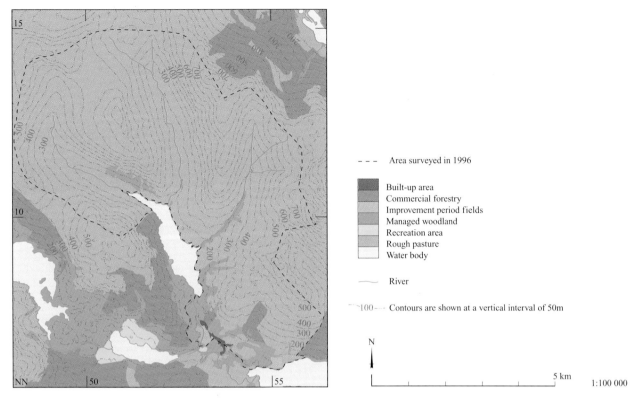

Map 9 Glen Finglas: current landuse (NMRS DC42595). This map is reproduced from Ordnance Survey material with the permission of Ordnance Survey on behalf of the Controller of Her Majesty's Office © Crown Copyright GD03127G001/2000.

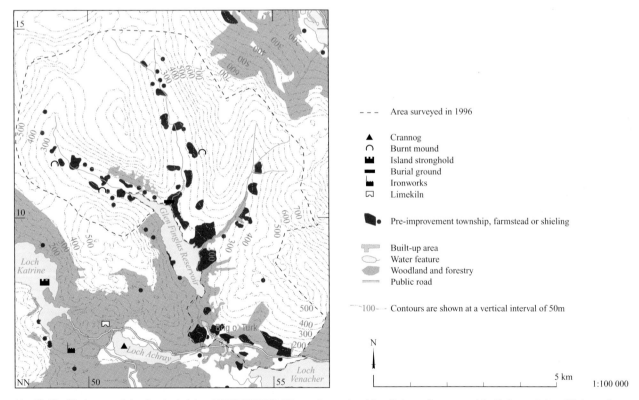

Map 10 Glen Finglas: recorded archaeological sites (NMRS DC42596). This map is reproduced from Ordnance Survey material with the permission of Ordnance Survey on behalf of the Controller of Her Majesty's Office © Crown Copyright GD03127G001/2000.

Flanders Moss

Below Aberfoyle, the valley of the River Forth opens out into a low-lying basin, about 5km in breadth, defined by the Menteith Hills to the north and the uplands of Kippen Muir to the south. During the most recent period of glaciation, which ended about 8000 BC, glaciers spread eastwards into this basin from Loch Lomond, and the limit of the ice sheet is marked by the Menteith Moraine, a gravel ridge running across the valley from Port of Menteith to Arnprior. Peat growth began after the retreat of the ice, but was halted as the sea level rose, depositing marine silts and clays. Subsequently, relative sea level fell once again and, from the fifth millennium BC, the area was colonised by extensive peat mosses.

The mosses, which reached depths of over 6m, were mostly cleared away in the eighteenth and nineteenth centuries in order to bring the fertile clays beneath into cultivation, and only fragments now survive, the largest being Flanders Moss East and Flanders Moss West (Map 11). So great has been the impact of drainage schemes and agricultural improvement that the former extent of the wetlands is barely reflected at all in the modern landuse pattern, with improved pasture running onto the hills to north and south of the valley. A reliable map of the former extent of the wetlands and peat mosses is not easily drawn, but the 30m OD contour can be used to indicate the extent of low-lying ground, and this has been highlighted on Maps 11 and 12 to frame the likely maximum area of wetlands. Drainage and land improvement is likely to have had a substantial impact on archaeological remains, both through direct destruction and through subsequent desiccation of organic deposits. Although some of the area shown on Maps 11 and 12 lies outwith the likely southern boundary of the Park, the whole area forms an integral cultural landscape and could be considered for inclusion.

During the Mesolithic and Neolithic periods the marshlands of the Forth would have provided a rich resource for early settlers. Stray finds of antler and stone implements recovered from the moss may have been lost during hunting expeditions, while a timber platform recently excavated at the edge of the moss to the north of Arnprior may have been built as a jetty. Radiocarbon dates indicate that the platform was constructed in the late fourth millennium BC (Ellis 1999, 214-8). Antiquarian records also mention the discovery of wooden trackways during the clearance of the mosses in the eighteenth century. Above the floor of the valley, especially in the hills to the north and west, there is more plentiful evidence for settlement at this time. The moorland to the north-west of the Lake of Menteith may have been a particular focus for ritual activity; here there is a cluster of at least fifty cup-and-ring marked boulders, a group of three burial cairns and a setting of standing stones, and there are more cup-markings to the south of Aberfoyle.

In later prehistory, settlement appears to have clustered around the edge of the moss. Fifteen prehistoric settlements are depicted on Map 12 – eight forts, three brochs, three homesteads and one dun. Apart from a homestead on Flanders Hill near the centre of the area, they are all located on the rising ground that fringes the valley, and a comparison with Map 11 shows that they are almost all situated on the better quality ground, now taken up by improved fields. Many of the

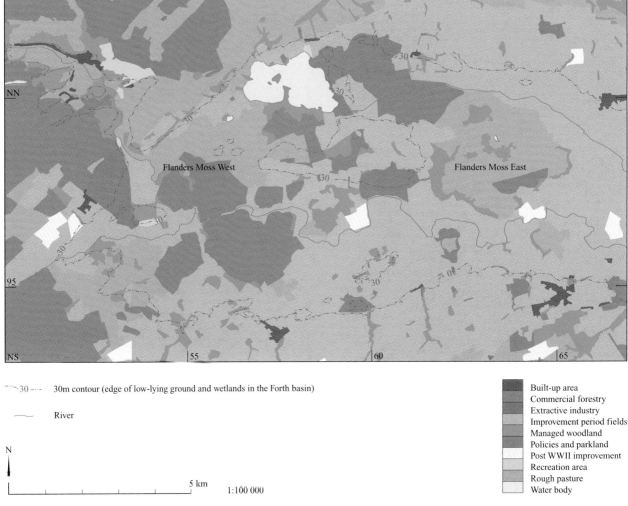

Map 11 Flanders Moss and the upper Forth valley: current landuse (NMRS DC42597). This map is reproduced from Ordnance Survey material with the permission of Ordnance Survey on behalf of the Controller of Her Majesty's Office © Crown Copyright GD03127G001/2000.

Turf houses, Carse of Stirling, drawn by Joseph Farington in 1792. The walls of houses such as these were carved out of the mosses in the Forth valley, and left free-standing as the peat around them was cleared away (© British Museum; top, NMRS ST576 & bottom, NMRS ST575).

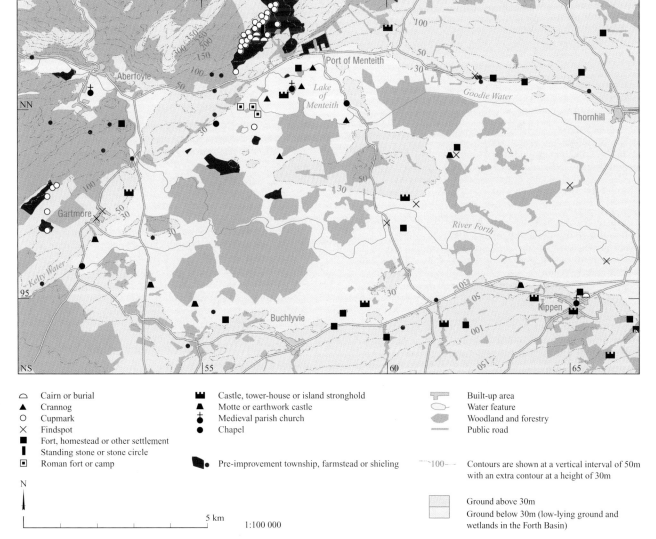

Map 12 Flanders Moss and the upper Forth valley: prehistoric, Roman and medieval monuments (NMRS DC42598). This map is reproduced from Ordnance Survey material with the permission of Ordnance Survey on behalf of the Controller of Her Majesty's Office © Crown Copyright GD03127G001/2000.

The policies of Cardross House are surrounded by improvement-period fields. Beyond is Flanders Moss East, one of the largest surviving mosses in the Forth valley (NMRS D56920CN).

best-preserved sites here have survived because they stand in pockets of woodland, such as the forts at Wester Arngibbon near Arnprior, or Keir Hill of Dasher near Kippen, while most of the sites in open locations have been ploughed down and are only known from antiquarian accounts (e.g. Keir Brae of Garden, near Arnprior) or as cropmarks (e.g. Portend homestead by the Lake of Menteith). It is likely that many more prehistoric monuments have been lost to cultivation.

A broadly similar pattern of settlement emerges for the medieval period, with most mottes, homestead moats and tower-houses concentrated on the better ground around the edges of the moss. These were the homes of the élite, but little is known of the pattern of vernacular settlement at this time. Extensive colonisation of the low ground by the middle of the eighteenth century is suggested by General Roy's map (Roy 1747-55), which depicts and names most of the farms that today occupy the valley floor, but it is not yet clear when these settlements were first established.

There are several substantial policies around the edges of the moss. In the main they owe their appearance to the radical developments of the improvement period, though some of them have developed around tower-houses, and probably have earlier origins. The policies at Gartmore, for example, around the ruins of Gartartan Castle were certainly well developed by the time of Roy's map. It is noteworthy that while many of the policies around Loch Lomond have been turned to recreational use, this has not happened in the upper Forth valley, where most policies continue to function as parks around private houses. Presumably, the attraction of a location adjacent to Loch Lomond has drawn developers there. It remains to be seen whether an increasing demand for leisure facilities will have a similar effect on Menteith.

Summary and Management Considerations

This case study shows how the nature and use of Flanders Moss has changed over time, a highly dynamic landscape that has been substantially altered by human activity. It also highlights the great potential of the moss to reveal further information, particularly in the form of organic remains, which tend not to survive elsewhere. This is vitally important for our understanding of the cultural heritage of the Park, for directing conservation measures and for informing education and interpretation initiatives.

At the same time, however, this information is vulnerable to desiccation, and consequent loss, if the environment of the moss is modified further. In the agricultural landscape around the moss, the historic landscape is again subject to pressures from continuing agricultural use, expansion of forestry and large-scale development, which can both further damage surviving evidence and impact considerably on the historical character of the local landscape.

CONCLUSION

This study shows how extensively the area of the proposed Park has been settled and exploited throughout the post-glacial period. Despite modern pressures from agriculture, afforestation, urbanisation and, most recently, developments for leisure, considerable evidence of the past survives. The character of the modern landscape was formed in the late-eighteenth and nineteenth centuries, but the study has demonstrated that the evidence for earlier periods is rich and complex. It shows that there is great diversity within the cultural heritage, and that the nature of the evidence varies within the different landscape areas of the Park – uplands, lowlands, wetlands and lochs. The use of each of these areas has changed over time, contributing to the character of the landscape today and highlighting its dynamic nature. Conservation strategies for the Park should seek to recognise this diversity and try to ensure that the historic character of the landscape is carried forward into the future.

The study also makes it clear that there is considerable potential for further research into the cultural heritage, both to recover new information and to add detail to that which is already known. The rich cultural heritage of the Park provides important opportunities for access, recreation and tourism, for education and for interpretation. Furthermore, it offers a significant focus for people, whether local communities or visitors, to engage with the landscape, providing a link with the past as the Park develops into the future.

Management Considerations

The character of the landscape of the Park has been substantially influenced by human activity over time. Successive periods of use have obscured or obliterated earlier patterns of settlement and associated landuse, but a considerable range of evidence still survives in the modern landscape, in both highland and lowland areas, to give an indication of the complex historical development of the area.

If the landscape is to reflect its historical development for the future, then evidence for past use requires specific conservation strategies. As the major threats to the cultural heritage stem from development and the range of modern landuse practices, their conservation is best achieved at a landscape scale, within an integrated approach to landscape management and development control. In addition, however, active management of specific sites or landscape features is required to ensure their survival in good condition.

There is scope for further research to discover new sites and recover more information from those already recorded. Field survey, whether from the air or on the ground, excavation, documentary research and palaeoecology all provide complementary evidence towards an enhanced understanding of the historical use of the landscape.

With care, the surviving evidence can be used to tell the story of the human use of the Park. It can provide direct opportunities for interpretation, recreation and tourism. It is also a vital resource for education at all levels, whether directed towards understanding the human past, landscape history or people's use of their environment through time.

Glen Buckie, looking north towards Balquhidder. The former shieling grounds at the head of the glen (in the foreground) have been planted with conifers; on lower ground extensive remains of pre-improvement settlement survive around the modern hill farms (NMRS D56959CN).

REFERENCES

Aitken, W G 1973
'Excavation of Bloomeries in Rannoch, Perthshire, and elsewhere'
Proc Soc Antiq Scot, 102 (1969-70), 188-204

BUFAU
Birmingham University Field Archaeology Unit

BUFAU 1997
Ben Lomond Archaeological Survey (Unpublished report, NMRS MS/975/1)

Campbell, R D 1999
Loch Lomond and the Trossachs
Edinburgh

Carter, S and Dalland, M 1997
Glenfinglas Estate: An Archaeological Survey
(Unpublished report, NMRS MS/899/23)

Dyson Bruce, L, Dixon, P, Hingley, R and Stevenson, J 1999
Historic Landuse Assessment (HLA): Development and Potential of a Technique for Assessing Historic Landuse Patterns
Historic Scotland Research Report
Edinburgh

Ellis, C 1999
Wetland Archaeology: Carse of Stirling Archaeological Assessment
(Unpublished report for Historic Scotland by AOC Archaeology Group

Fairhurst, H 1971
'The Deserted Settlement at Lix, West Perthshire'
Proc Soc Antiq Scot 101 (1968-69), 160-99

FIRAT 1995-8
Loch Lomond Islands Survey
(Unpublished reports, NMRS MS/993/1-5)

Henderson, J C 1999
'A survey of crannogs in the Lake of Menteith, Stirlingshire'
Proc Soc Antiq Scot, 128 (1998), 273- 92

Hunter, J R 1996
The Ben Lomond Project. Report on the 1995 season
(Unpublished report, NMRS MS/969/2)

Main, L 1999
'Excavation of a timber round-house and broch at the Fairy Knowe, Buchlyvie, Stirlingshire, 1975-8' *Proc Soc Antiq Scot,* 128 (1998), 293-417

RCAHMS
Royal Commission on the Ancient and Historical Monuments of Scotland

RCAHMS 1963
Stirlingshire: An Inventory of the Ancient Monuments
Edinburgh

RCAHMS 1978
The Archaeological Sites and Monuments of Dumbarton District, Clydebank District, Bearsden and Milngavie District, Strathclyde Region
Edinburgh

RCAHMS 1979
The Archaeological Sites and Monuments of Stirling District, Central Region
Edinburgh

Roy, W 1747-55
Military Survey of Scotland

Smout, T C 1997 (ed.)
Scottish Woodland History
Edinburgh

Stewart, J H and Stewart, M B 1989
'A highland longhouse - Lianach, Balquhidder, Perthshire'
Proc Soc Antiq Scot, 118 (1988), 301-317

SNH 1999
National Parks for Scotland; Scottish Natural Heritage's Advice to Government
Battleby

Further Reading on Conservation of the Cultural Heritage

Historic Scotland *Memorandum of Guidance on listed buildings and conservation areas.*

Historic Scotland *Technical Advice Notes:* series of publications on various aspects of conservation of the built heritage.

Historic Scotland *The Historic Scotland Guide to Conservation Plans.*

Historic Scotland *The Stirling Charter: Conserving Scotland's Built Heritage.*

Scottish Executive Development Department *National Planning Policy Guideline 5: Archaeology and Planning.*

Scottish Executive Development Department *Planning Advice Note 42: Archaeology.*

Leaflets Available Free from Historic Scotland

A Guide to Conservation Areas in Scotland.

Archaeological Information and Advice in Scotland.

Archaeology on the Farm (joint publication with the Council for Scottish Archaeology).

Farm Architecture: The Listing of Farm Buildings.

Grants for Ancient Monuments.

Managing Scotland's Archaeological Heritage.

Scheduled Ancient Monuments: a guide for owners, occupiers and land managers.

Scheduled Ancient Monuments and Metal Detectors.

Scotland's Listed Buildings: A Guide for Owners and Occupiers.

The Carved Stones of Scotland.

Contact Addresses

Mrs L Main
Archaeology Officer
Environmental Services
Stirling Council
Viewforth
STIRLING
FK8 2ET
Tel: 01786-442752
Email: mainl@stirling.gov.uk

Mr R Turner
Archaeologist
The National Trust for Scotland
28 Charlotte Square
EDINBURGH
EH2 4ET
Tel: 0131-243 9300
Email: rturner@nts.org.uk

West of Scotland Archaeological Service
Charing Cross Complex
20 India Street
GLASGOW
G2 4PF
Tel: 0141-287 8333
Email: archaeology.wosas@virgin.net

Mr T Yarnell
Archaeologist
Forestry Commission
231 Corstorphine Road
EDINBURGH
EH12 7AT
Tel: 0131-334 0303
Email: tim.arch@forestry.gov.uk